Fifth Dimension

Fifth Dimension

The Light to See

Marc E. King

To order additional copies of this book, contact:
Xlibris Corporation
1-888-795-4274
www.Xlibris.com
Orders@Xlibris.com
121506

Contents

Introduction for the Technical Version...................................7

Summary of Results

    From Text and Appendices of Changing Your Mind
    A Theory of Space without Time 9

Appendix P

    Progression of the Spatial Sequence 13

Appendix Q

    The Magnetic Force 17

Appendix R

    The Periodic Chart of Elements 27

Appendix S

    Derivative Coordinates for Two Added Vertices 31

Conclusions for Five Dimensions 43

Appendices for Reference from the Original Manuscript:

Appendix C (Reference)

A General Fibonacci Calculation 45

Appendix K (Reference)

The Various Sizes of Black Holes and Curvatures of Three-Dimensional Space 47

Appendix L (Reference)

Definition of Mass and Geometry for Black Holes 49

Appendix N (Reference)

Nuclear Forces 53

References: 55

Introduction for the Technical Version

This text adds Appendices P, Q, R, and S to the existing appendices to the manuscript *A Mathematical Transformation of Variables Defining Space-Time and the Constant h*.

The new appendices assert the charge and mass relationships regarding chemistry, nuclear physics, and the related force fields known as **E** and **B**.

The origin of light is presented in a spatial (non-temporal) view.

Summary of Results

From Text and Appendices of
Changing Your Mind
A Theory of Space without Time[1]

Per the main text and manuscript, we assert the transformation
$t = cB$

where t (sec) = c (meter/sec) x B (sec^2 / meter) = c / b (event/ meter) and where a single event = one sec^2 and where we need to remove "time" t from our units of measure for non-macroscopic physics.

It has been shown that

$b = 1.111E\text{-}17$ (meters), and that

$E_B = 680eV / 1kg / (b^3 met^3)$.

The Fibonacci infinite sequence can be described as follows:

Lim (n→infinity) F(n-1) / F(n) = φ = 0.618 . . . and

Lim (n→infinity) F(n-2) / F(n) = γ = 0.382 . . . .

The growth rate of space itself for three dimensions expanding through five dimensions is

r_V = φ^(5-3) so that

V_5 / V_3 = e^(r_V) for one "second" of time t.

Similarly, the growth rate of the natural logarithmic base is

r_L = (1/γ)^(5/2) so that

e→e^(r_L) for D=5 traversing through D=8.

Appendix C is attached to this text for reference.
Quantum mechanics is then described as E / E_B = κc / n and

h = fn(E_B) = b$E_B$$\kappa$.

The definition of quantum mechanics becomes mass x volume and defines the chemistry of the periodic chart of elements, and the Schrödinger[2] solution is replaced by

E' = E_B(1 – p/n²).

Dimensional spatial curvatures are defined as $0 \le c \le 1$ with the terms

0 = open, and

1 = closed.

Appendix K is attached to this text as a reference for curvatures.

Appendices L and N are attached as references for 5-dimensional geometry and 5-dimensional nuclear forces respectively.

The sequence of spatial progression can be described by the following Appendix P.

Appendix P

Progression of the Spatial Sequence

The concept of continuous "time" along with the reality of spatial progression is represented in the diagram <u>One Dimensional Spatial Progression Model</u>.

While the sequence moves directionally in three dimensions through five dimensions (the three-five-three vertical arrows) we perceive only the 3-dimensional progression (D = 3 horizontal arrows,) due to our limited synaptic "speed" or rate-of-perception.

The "speed of light" (denoted c) is known to be 300,000,000 meters per "second" of our perceived time t. The speed of light is much greater than the fastest space ship or satellite velocity and represents the distance b (one second of perceived time divided by 300,000,000 or a tiny fraction of one second.) The metric b is shown in the following diagram of sequential spatial progression.

The distance b can be thought of as the smallest or most precise possible movement through space. The distance b can equally be thought of as the smallest unit of measure on a ruler or other measuring stick, e.g. as one millimeter is a small segment of one centimeter and a very small segment on a meter-long measurement stick, and as one mil (thousandths of an inch) cannot even be discerned on a yard-stick segmented by increments of feet, inches, quarter inches, eighth inches, and sixteenth inches.

While the extreme value of the speed of light c allows for a very small increment along a one dimensional "meter measurement stick" (the value b,) our vision and our entire neurological and synaptic ability allow only for the minimum perceivable increment of about 1000 x b. For example, we can see the inch line on a ruler but we cannot see the one-mil "line."

In other words, we do not have and never will have the ability to perceive the higher dimensions even though we exist arguably 50% in dimension D=5 and above and the other 50% in our perceivable dimension D=3.

While our minds understand the fifth dimension (memories, analytical thought, dreams and more,) our sensory perceptions do not allow a clear and visualized understanding of higher dimensions.

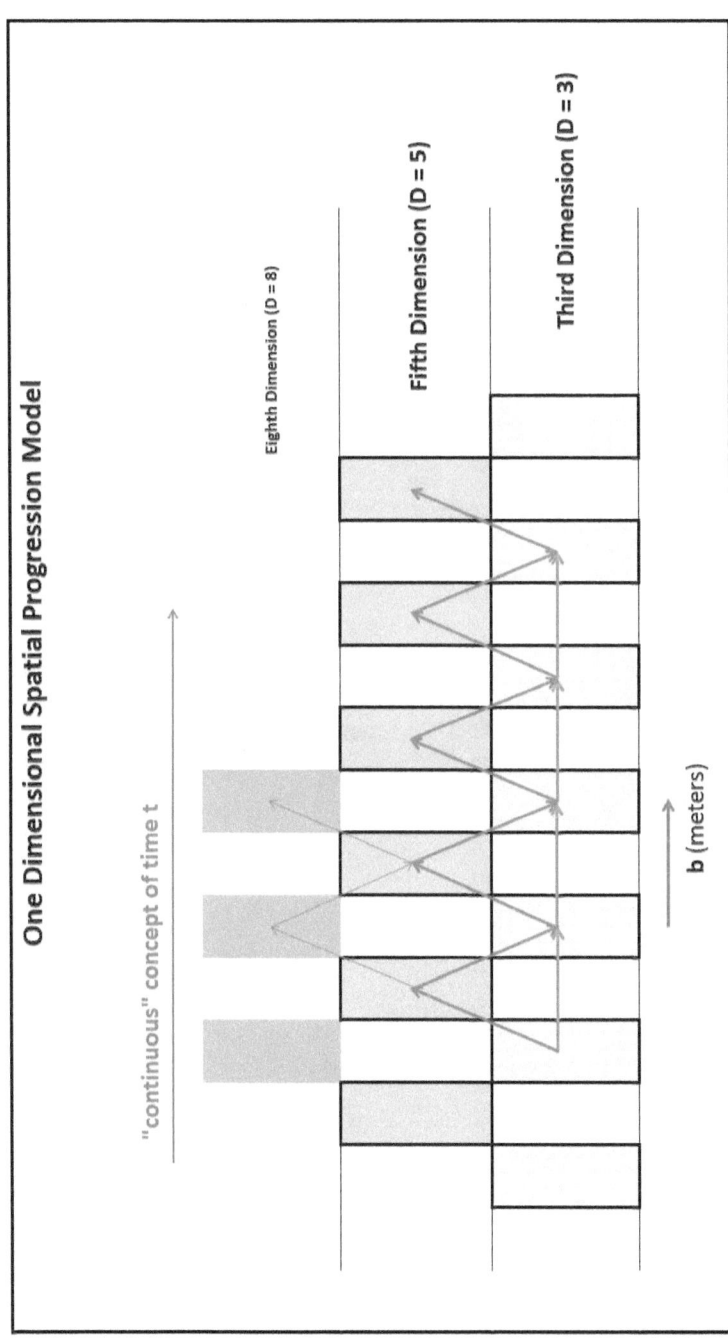

Appendix Q

The Magnetic Force

Why Does $\mathbf{F}_M = q\mathbf{v} \times \mathbf{B}$?

When the author was an undergraduate, he asked his most respected physics professor why $\mathbf{F}_M = q\mathbf{v} \times \mathbf{B}$?

The professor answered "because the force obeys the right-hand rule, you must know that since you have worked all the problems correctly."

The author replied, "I know that, but I don't understand why the force
$\mathbf{F}_M = q\mathbf{v} \times \mathbf{B}$ exists or, for that matter, why it obeys a right-hand rule instead of a left-hand rule?"

The professor replied "Well! If that's what you are asking, then there are only two possible answers. One is God and the other is random chance. You will need to decide for yourself."

Here is the correct answer:

$\mathbf{F}_M = q\mathbf{v} \times \mathbf{B}'$ as a singe-plane right-angle example becomes $F_M = qvB'$ where we need to alter the standard nomenclature since we have assigned the letters B and b to represent the transformation $t = cB = c(1/b)$.

Then

$F_M / qb = B' / cB = bB' / c$ or

$F_M / qb^2 = B' / c$

where the vector $\mathbf{F}_M$ has the same direction as the sequential spatial progression and is orthogonal to the surface b^2 while

the vector $\mathbf{B}'$ has the direction dictated by the attributes + and –.

Therefore, $\mathbf{F}_M = q\mathbf{v} \times \mathbf{B}'$ using the "right" hand rule.

Q.E.D.

How and/or where is the surface being pushed or pulled in the "forward" sequential progression?

Through the next (5th) dimension in order to achieve the next sequential 3-dimensional frame at the sequential rate c.

Since the force acts "throughout a two-dimensional surface" instead of "along a one-dimensional line," then the force is based in five dimensional space.

The effective three dimensional force $\mathbf{F}_M$ is then described by the 3-dimensional vector $\mathbf{B}'$.

Then the magnetic force $\mathbf{F}_M$ should provide information about a via-way or possible vector map from 3 to 5 dimensional space.

Boundary conditions:

1. $F_{Q5} = Q_0 \, f(q_1/\text{met}, q_2/\text{met}) / r^3$

2. **Projection** (D=3) of $F_{Q5} = B' / c$.

These lead to the three-dimensional observation of an effective "charged-point" moving in a closed linear (circular) manner in a plane orthogonal to the 3-dimensional force F_M.

In fact, this observation should be a three-dimensional view of closed charged lines "moving" on a closed surface (spherical shell) about a three-dimensional volume = $4/3 \, \pi \, r^3$ where B' is aligned with r.

Per Appendix N, the atomic nucleus is not 3-dimensional (breaks the 3-dimensional mass x volume rules) and is at least 5-dimensional (having the value of $b = b_5 = b_3 / c$ per Appendix L.)

Using a chosen spherical model for the He nucleus, the force pushing (compressing) the two positive charges together:

$$F_N = F_{Q5} = Q_0 \, (e+ / 2\pi r_N)^2 / r_N^3$$

and is extremely large by our three-dimensional experience.

3. Boundary **Gradient**

The boundary gradient defines so-called discontinuity. Since there is no real continuity, there is also no real discontinuity.

For the planar surface between three-dimensions and five-dimensions, the value b must traverse from b_3 to b_5 or from 1.111E-17 meters to 3.703E-26 meters in the increment $b_8 = 4.114\text{E-}43$ meters.

Then the intersection gradient of $b = \Delta b / \Delta x = 1.111\text{E-}17 / 4.114\text{E-}43$

$= 2.70\text{E+}25 = c^3$ and represents the rate of spatial change (the effective change in the "size" of space) across the boundary b_g.

Then the energy gradient $\text{GradE}_B = c^3$

where GradE_B has the units J / meter / Kg-$b_3{}^2$

= Intersection Force per Mass-Boundary.

We can now write an expression for our prior He model of nuclear force:

$\mathbf{F}_N = \mathbf{F}_{Q5} = \mathbf{Q}_0 \, (e{+} / 2\pi r_N)^2 / r_N{}^3$ or

$\text{GradE}_B = \text{mag}(\mathbf{Q}_0) \, (e{+} / 2\pi)^2 / r_N{}^5$. Then

$\mathbf{Q}_0 / r_N{}^5 = \text{GradE}_B / (e{+} / 2\pi)^2$ for the He nucleus, and

$\mathbf{Q}_0 = (\, 0, \, 0, \, \text{GradE}_B \times r_N{}^5 / (e{+} / 2\pi)^2, \, 0, \, 0 \,)$.

Then $qv = (\, 1q \text{ met} / c, \, 0, \, 0 \,)$ and $\mathbf{B'} = (\, 0, \, B', \, 0 \,)$

and $\mathbf{F}_M = qv \times \mathbf{B'} = (\, 1q \text{ met} / c, \, 0, \, 0 \,) \times (\, 0, \, B', \, 0 \,) = (\, 0, \, 0, \, F_M \,)$.

Then $\text{Mag}(\mathbf{F}_M) = F_M = B' / c$ and

$B' = c \times \text{GradE}_B \times r_N{}^5 / (e{+} / 2\pi)^2 = c \, Q_0$ with units (mass / q) event$^{-1/2}$.

Or we can write:

$B' = c Q_0 \, (\text{mass} / q) \times (1 / cB)$ and

$B' = bQ_0$ with units (mass / q) per frame, or

$B' = Q_0 (r_N / b)$ with units mass / q. Equivalently,

Charge / Mass = $(1/ B') = b / (Q_0 x r_N)$ and it follows:

r_N = (Mass / Charge) b / Q_0 = (Mass / Charge) b $(e+ / 2\pi)^2$ / $(GradE_B x r_N^5)$,

$r_N^6 = b$ (Mass / Charge) $(e+ / 2\pi)^2 / c^3$, and for a considered
2-amu *He*

$r_N = (b x 2amu/2e+ x (e+ / 2\pi)^2 / c^3) \wedge 1/6$, or

r_N = **3.071E-14 meters in three dimensions.**

For **_4-amu_** (stable) He, then

$r_N \rightarrow r_N x 2^{(1/6)}$ = **3.447E-14 meters.**

Returning to the vector projection from 5 to 3 dimensional space, we see that x and y are curved in 5-dimensional space (having curvatures 1) while z remains orthogonal to the surface xy (having surface curvature 1) and points along the radius defined by the center of closed geometries. z appears to have curvature 0.

It is possible the 4th directional value for the 5th dimensional vector has a direction tangent to the closed surface xy and that the 5th directional value has a tangential direction that is orthogonal to tangent 4. Then these should represent multi-dimensional motion of the tangent "lines" or "axis pair."

These lines (axes) should have curvature 0. Then:

Possible vector $\mathbf{Q_5} = Q_5 (x_c, y_c, z_r, t_x, t_y)$ or using other nomenclature,

Possible vector $\mathbf{Q}_5 = Q_5 \, (x_T, y_T, z_R, \textit{arc } z_X, \textit{arc } z_Y)$,

and t_x and t_y (derivatives) represent a change-of-spatial-direction at the three-dimensional point (x_c, y_c, z_r), and are orthogonal to x_c and y_c.

One way to visualize 5-dimensional coordinates is shown by the following diagram. Another way is to view the inverted diagram.

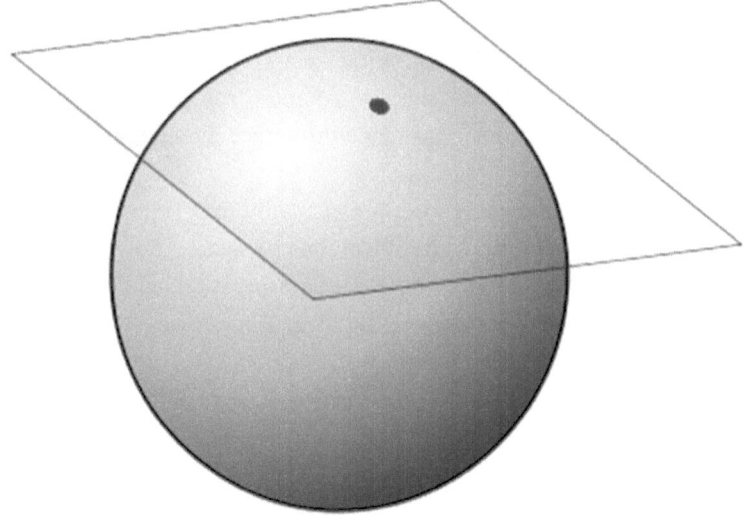

Effectively, the planar system of coordinates (t_x, t_y) moves as a function of (x_c, y_c, z_r).

Perhaps the best way to visualize three dimensional space "moving" or progressing through five dimensional space is to turn the above diagram upside down and view the inverse image (a ball sitting on a table top.)

Next, imagine the ball moving on the surface with the curvature 0 (a straight line) and with the curvature 1 (a circle where the ball would return to it's starting position.)

Now, imagine the ball moving in a spiral with curvature φ (the Fibonacci spiral) and this view should represent the spatial progression of 3 dimensional space through 5 dimensional space.

This view makes it more clear why our memories can be addressed through five-dimensional space. The two extra coordinates (or axes) provide the exact location and a way to address (access) any pre-sequential three-dimensional "frame."

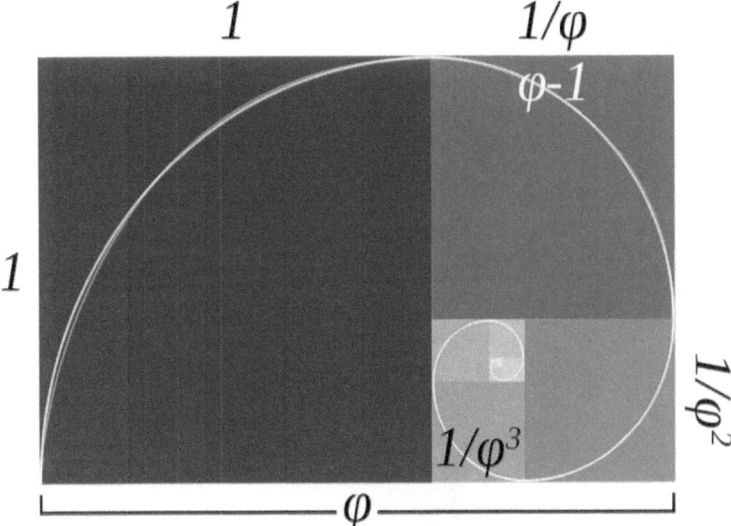

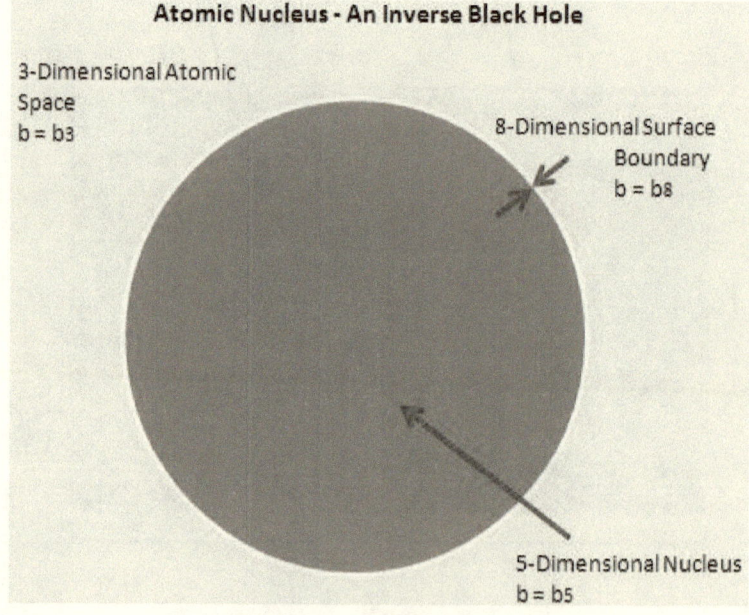

Atomic Nucleus - An Inverse Black Hole

3-Dimensional Atomic Space
$b = b_3$

8-Dimensional Surface Boundary
$b = b_8$

5-Dimensional Nucleus
$b = b_5$

Appendix R

The Periodic Chart of Elements

Then there are two boundary conditions (per Appendix E and Appendix Q) defining the natural elements:

1. E / E_B = mass x volume = atomic mass x $4/3 \, \pi \, r_{max}^3$

2. r_N = (nuclear mass / nuclear charge) x (b / Q_0)

where r_{max} is the least bound (closest to E_B) electron atomic energy state radius, and

where r_N is the nuclear radius.

The atom and nucleus must obey both conditions.

While the nucleus achieves a lower bound energy requirement with increasing mass, the atomic mass remains bounded by mass x volume.

Group → / Period ↓	1	2	3	4	5	6	7	8	9	10	11	12	13	14	15	16	17	18
1	1 H																	2 He
2	3 Li	4 Be											5 B	6 C	7 N	8 O	9 F	10 Ne
3	11 Na	12 Mg											13 Al	14 Si	15 P	16 S	17 Cl	18 Ar
4	19 K	20 Ca	21 Sc	22 Ti	23 V	24 Cr	25 Mn	26 Fe	27 Co	28 Ni	29 Cu	30 Zn	31 Ga	32 Ge	33 As	34 Se	35 Br	36 Kr
5	37 Rb	38 Sr	39 Y	40 Zr	41 Nb	42 Mo	43 Tc	44 Ru	45 Rh	46 Pd	47 Ag	48 Cd	49 In	50 Sn	51 Sb	52 Te	53 I	54 Xe
6	55 Cs	56 Ba		72 Hf	73 Ta	74 W	75 Re	76 Os	77 Ir	78 Pt	79 Au	80 Hg	81 Tl	82 Pb	83 Bi	84 Po	85 At	86 Rn
7	87 Fr	88 Ra		104 Rf	105 Db	106 Sg	107 Bh	108 Hs	109 Mt	110 Ds	111 Rg	112 Cn	113 Uut	114 Fl	115 Uup	116 Lv	117 Uus	118 Uuo

Lanthanides	57 La	58 Ce	59 Pr	60 Nd	61 Pm	62 Sm	63 Eu	64 Gd	65 Tb	66 Dy	67 Ho	68 Er	69 Tm	70 Yb	71 Lu
Actinides	89 Ac	90 Th	91 Pa	92 U	93 Np	94 Pu	95 Am	96 Cm	97 Bk	98 Cf	99 Es	100 Fm	101 Md	102 No	103 Lr

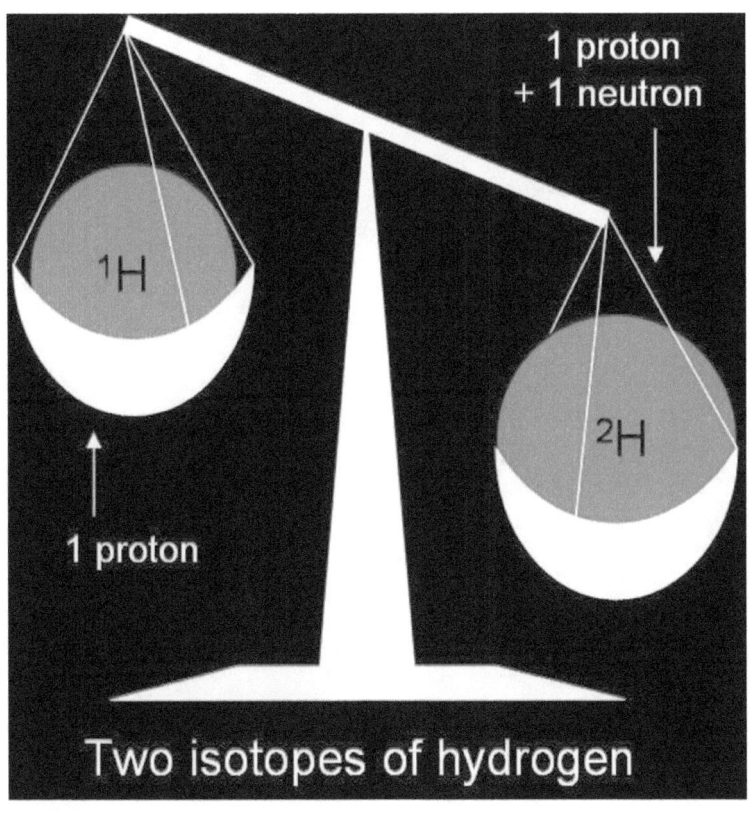

Two isotopes of hydrogen

Appendix S

Derivative Coordinates for
Two Added Vertices

A derivative, e.g. $df(x)/dx$ is tangent to the curve $f(x)$ similar to the 5-dimensional coordinates t_x and t_y from Appendix Q.

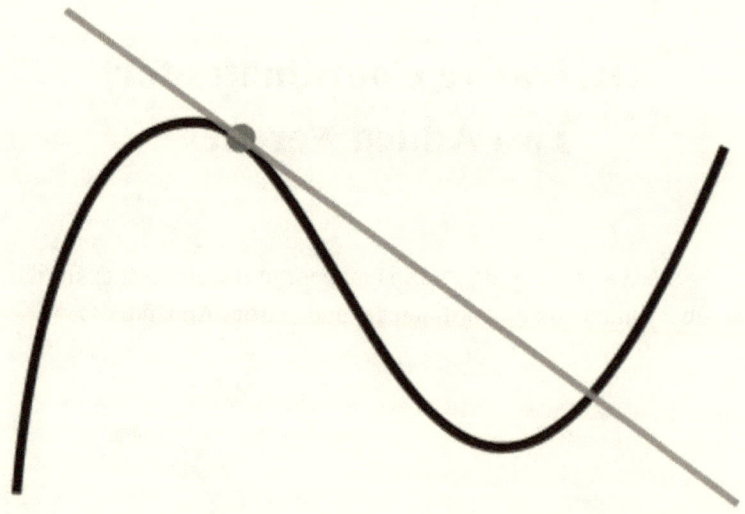

The derivative itself is independent from (orthogonal to) the point x, f(x).

The derivative is a function at the point x, f(x) relative to the points following and preceding the point x, f(x). It indicates the direction from the existing position to the next position.

In the Fibonacci spatial progression, we are traveling from the spatial dimension D=3 to D=5 and from the power-of-physical-events 2 to 3 as in the accelerations from 3 and 5 dimensional forces respectively.

The $\lim (n \rightarrow \text{infinity})\ F(n-1) / F(n) = \varphi = 0.618$,

and this is the dimensional sequential growth rate. Then,

$dt_x/dx_c = dt_y/dy_c = \varphi$

and this matches the alternate view provided in Appendix Q.

The growth rates from Appendix C can now be more easily visualized.

To understand the rate of spatial growth using dimension 3 traversing through dimension 5 as an example:

$r_V = \varphi^{\wedge}(5\text{-}3) = \varphi^{\wedge}2$ or φ^2

meaning we have now added two extra dimensions (vertices) each having the growth rate φ so that the rate of additional volume growth is φ^2.

To understand the increasing value for 3-dimensional $e = 2.718$ (the base of natural logarithms) in higher dimensions D=5 and above, we need to review the dimensional geometry as follows:

Note: **Here we apologize for confusion in nomenclature regarding the Fibonacci ratio and the Euler[3] angle both known as φ. Within the scope of this appendix, these will not be equated.**

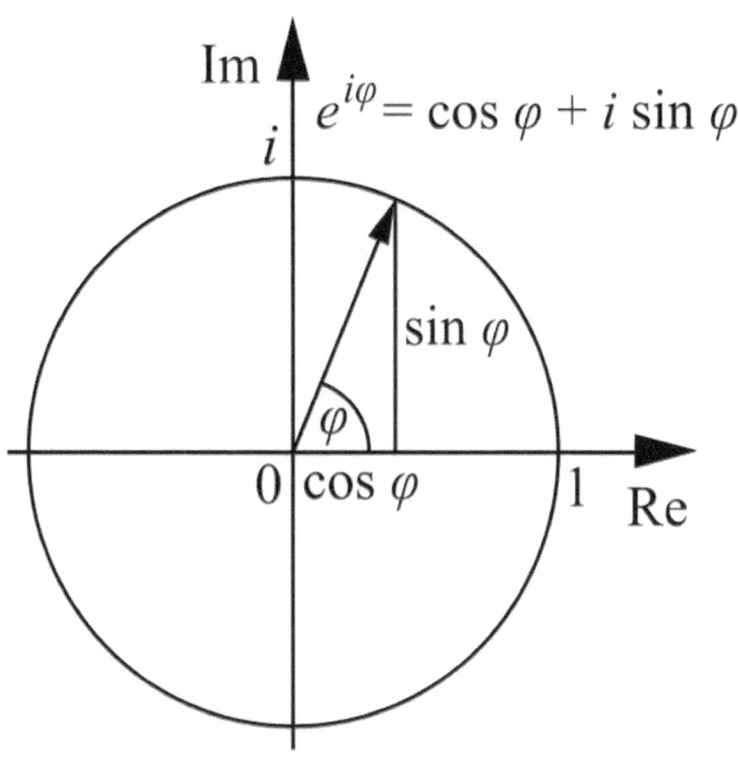

In a similar way to the rate of volume expansion and per Appendix C, the expansion of the logarithmic base (growth rate of the radius R in the above diagram) represents the addition of extra vertices. The number of vertices has grown from 3 to 3+2 = 5. Then,

$$r_L = (1/\gamma)\char94 5/2$$

using D=5 traversing through D=8 as an example and where γ is defined in Appendix C:

$$\gamma = \lim (n \rightarrow \text{infinity}) \, F(n-2) / F(n).$$

The physical concept

$$i = (-1)\char94 1/2 \text{ or } (-1)^{1/2}$$

is then represented by the example vector:

$$\mathbf{R} = (0,0,0,1,0).$$

We can now physically extrapolate for charged points, lines, surfaces and volumes as well as their relative motions:

$\mathbf{E}_3$ is orthogonal to $\mathbf{B}_2$ where:

$\mathbf{E}_3$, $\mathbf{B}_2$, and $\mathbf{EB}_5$ represent open geometries moving within closed geometries of the same dimension, i.e. a radius within a circle and an open plane within a spherical surface respectively for $\mathbf{E}$ and $\mathbf{B}$.

We can express sequential values for $\mathbf{E}$ and $\mathbf{B}$ as below:

E:	UP	0	DN	0
B:	0	DN	0	UP

And the corresponding vector sequence:

$$EB = EM = (0,0,1,0,0)\ (0,0,0,1,0)\ (0,0,-1,0,0)\ (0,0,0,0,1)$$

where the rate of vector progression through higher dimensional space is 1/4 cycle within b/4 meter and one complete cycle within the spatial progression b meters.

This is the nature of light.

The following diagrams represent achievable technology from the outline in appendices P through S.

These diagrams represent, among other things, the 5-dimensional computer model. The example atomic nucleus is Helium due to its multi-charge simplicity.

The smallest three dimensional energy ratio $E\ /\ E_B$ is more than enough to provide spin state motion for the positive charges.

For example, one so-called electron-spin-split energy state transition (fine state transition $2p_{3/2}$ to $2p_{1/2}$) for Hydrogen easily provides for the real nuclear "spin" difference having the value $E = 4.5E-5$ eV with the wavelength 2.7 cm.

Energies for charge motion have already been achieved through techniques such as magnetic resonance[4].

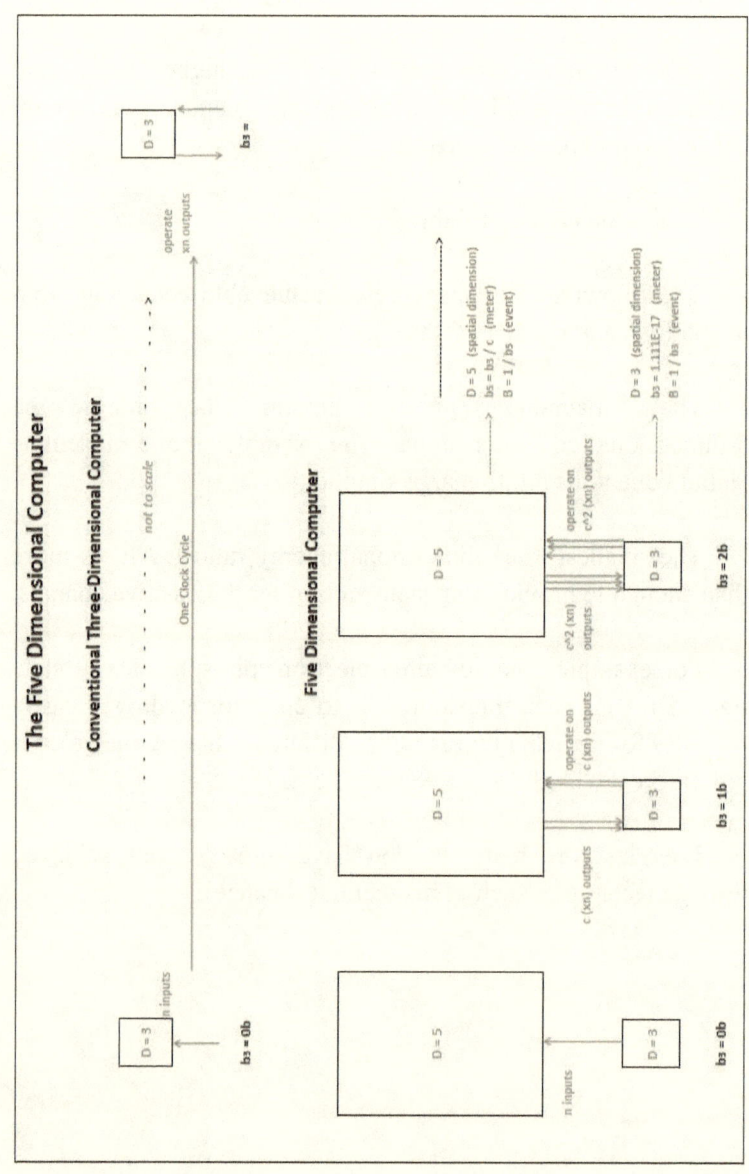

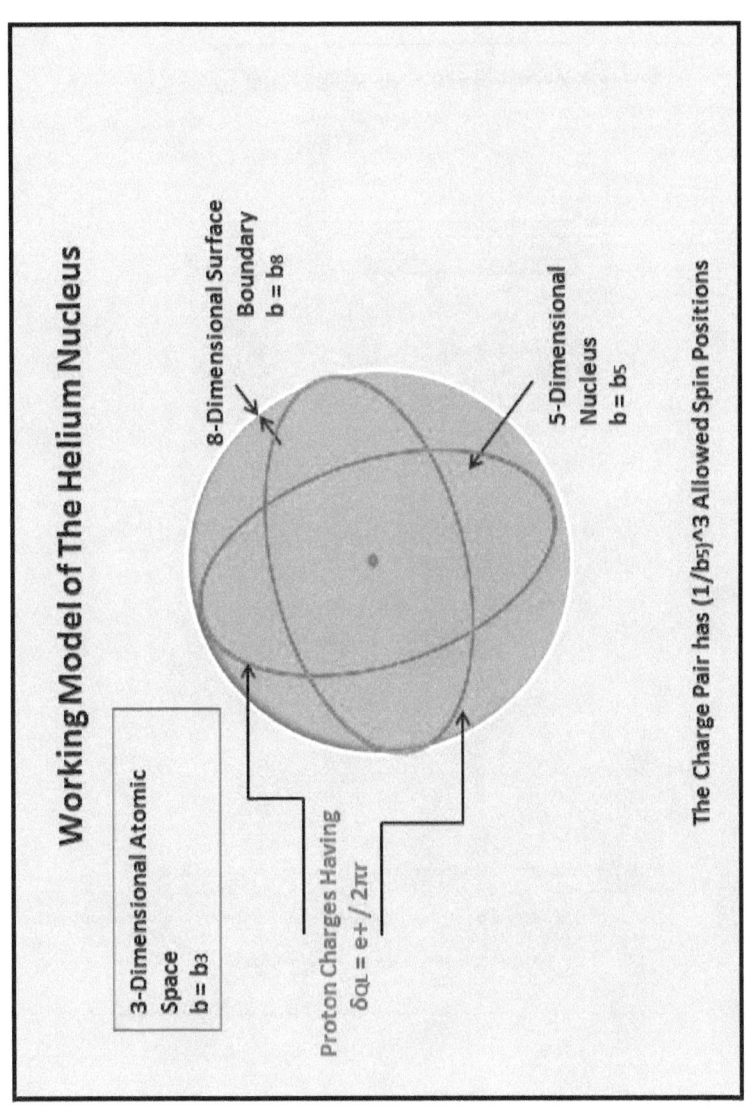

Working Model of The Helium Nucleus

8-Dimensional Surface
Boundary
$b = b_8$

5-Dimensional
Nucleus
$b = b_5$

3-Dimensional Atomic
Space
$b = b_3$

Proton Charges Having
$\delta_{QL} = e+ / 2\pi r$

The Charge Pair has $(1/b_5)^3$ Allowed Spin Positions

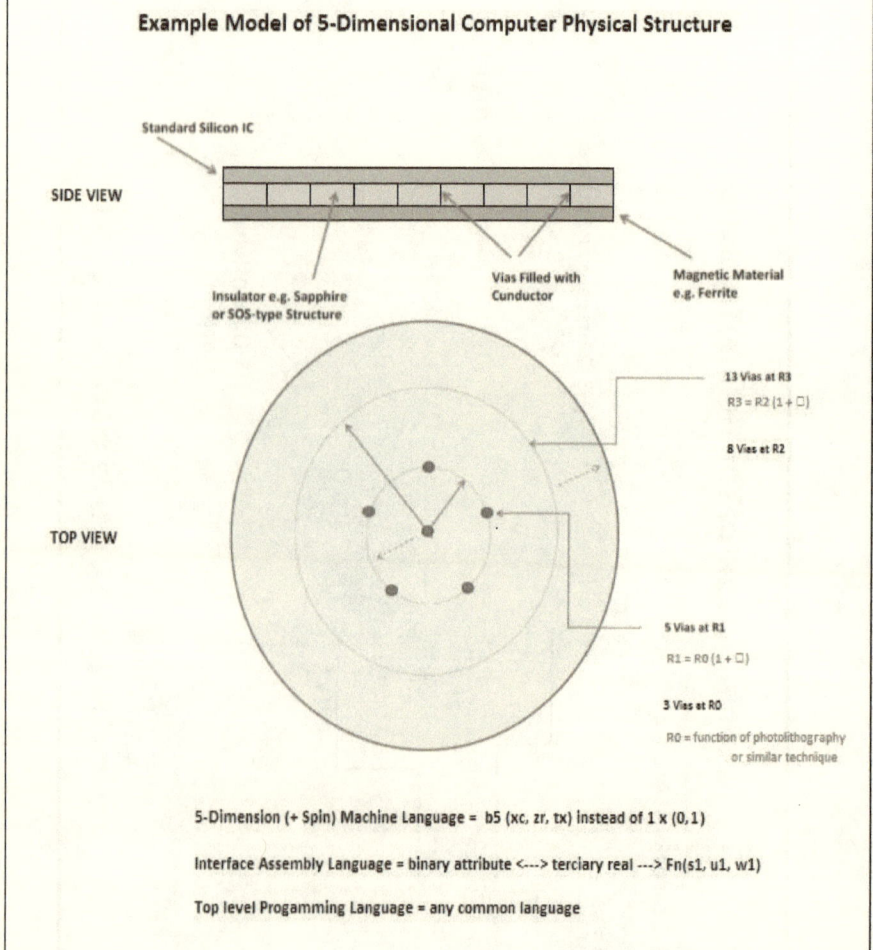

Example Model of 5-Dimensional Computer Physical Structure

SIDE VIEW

Standard Silicon IC

Insulator e.g. Sapphire or SOS-type Structure

Vias Filled with Cunductor

Magnetic Material e.g. Ferrite

TOP VIEW

13 Vias at R3

$R3 = R2(1 + \square)$

8 Vias at R2

5 Vias at R1

$R1 = R0(1 + \square)$

3 Vias at R0

R0 = function of photolithography or similar technique

5-Dimension (+ Spin) Machine Language = b5 (xc, zr, tx) instead of 1 x (0,1)

Interface Assembly Language = binary attribute <---> terciary real ---> Fn(s1, u1, w1)

Top level Progamming Language = any common language

OR- *FUNCTION* LOGIC GATE EXAMPLE (Binary Only)

Three Dimensional Logic

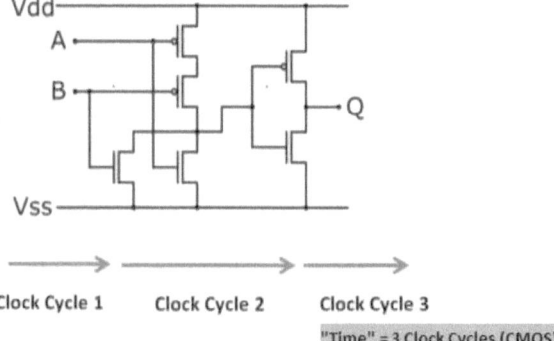

Clock Cycle 1   Clock Cycle 2   Clock Cycle 3

"Time" = 3 Clock Cycles (CMOS)

Five Dimensional Logic

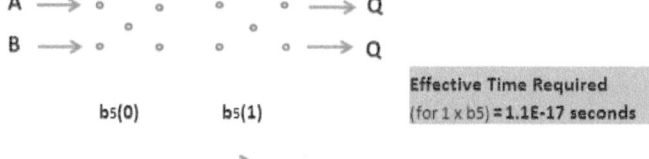

$b_5(0)$ $b_5(1)$

Effective Time Required
(for 1 x b5) = 1.1E-17 seconds

Example Bit Map for Compiler

word 1 / word 2 — clock 0sec

clock	bit 1	bit 2	bit 3	bit 4	bit 5	bit 6	bit 7	bit 8	bit 1	bit 2	bit 3	bit 4	bit 5	bit 6	bit 7	bit 8
0sec (b3=1, b5)	1	0	0	0	0	0	0	1	0	0	0	0	1	0	0	0
1	s1	s0	s0	s0	s0	s0	s0	s0	s1	s0	s0	s0	s0	s0	s0	s0
2	s2	s1	s0	s0	s0	s0	s0	s0	s2	s1	s0	s0	s0	s0	s0	s0
3	s3	s2	s1	s0	s0	s0	s0	s0	s3	s2	s1	s0	s0	s0	s0	s0
4	s4	s3	s2	s1	s0	s0	s0	s0	s4	s3	s2	s1	s0	s0	s0	s0
5	s5	s4	s3	s2	s1	s0	s0	s0	s5	s4	s3	s2	s1	s0	s0	s0
6	s6	s5	s4	s3	s2	s1	s0	s0	s6	s5	s4	s3	s2	s1	s0	s0
7	s7	s6	s5	s4	s3	s2	s1	s0	s7	s6	s5	s4	s3	s2	s1	s0
8	⋮	s7	s6	s5	s4	s3	s2	s1		s7	s6	s5	s4	s3	s2	s1
9	⋮			s6	s5	s4	s3	s2				s6	s5	s4	s3	s2
10	⋮					s5	s4	s3						s5	s4	s3

clock 3.3E-09sec

clock	bit 1	bit 2	bit 3	bit 4	bit 5	bit 6	bit 7	bit 8	bit 1	bit 2	bit 3	bit 4	bit 5	bit 6	bit 7	bit 8
	sn-0	sn-1	sn-2	sn-3	sn-0	sn-1	sn-2	sn-3	sn-1	sn-2	sn-3	sn-1	sn-0	sn-1	sn-2	sn-3
	sc-0	sc-1	sc-2	sc-3	sc-0	sc-1	sc-2	sc-3	sc-1	sc-2	sc-3	sc-1	sc-0	sc-1	sc-2	sc-3
3.3E-09sec (b3=2)	A-0	A-1	A-2	A-3	A-0	A-1	A-2	A-3	A-1	A-2	A-3	A-1	A-0	A-1	A-2	A-3

word 1   word 2

A(x) = Compiler Operator where A(x) = A(0, 1)

Conclusions for Five Dimensions

1. The atomic nucleus and its forces are 5-dimensional.

2. New mass-volume-charge quantum relationships determine the periodic chart of natural elements.

3. Electromagnetic radiation is the result of spatial progression at the rate c in three dimensions.

4. Vector mapping between 3 and 5 dimensional space is best described in the diagram <u>Coordinates for the Three Dimensional Sequence</u>.

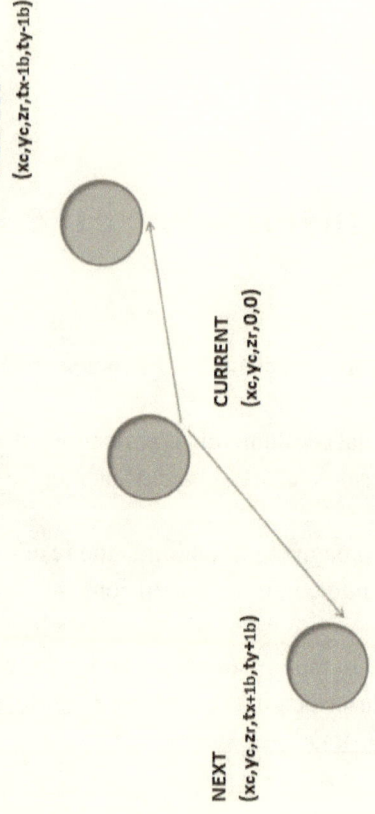

Coordinates for the 3-Dimensional Sequence

Three Dimensions Moving about Two Dimensions in Five Dimensional Space

PREVIOUS
$(x_c, y_c, z_r, t_x{-}1b, t_y{-}1b)$

CURRENT
$(x_c, y_c, z_r, 0, 0)$

NEXT
$(x_c, y_c, z_r, t_x{+}1b, t_y{+}1b)$

Appendix C (Reference)

A General Fibonacci Calculation

The Fibonacci infinite sequence was referenced in Appendix A,

$F(n) = F(n-1) + F(n-2)$ with seed values $F(0) = 0$ and $F(1) = 1$.

Ratios converge, and

$\lim(n \rightarrow \text{infinity})\ F(n+1) / F(n) = \varphi = (1 + 5^{\wedge}1/2) / 2 = .618 \ldots$
and

$\lim(n \rightarrow \text{infinity})\ F(n-2) / F(n) = \gamma = .382 \ldots$ and so on.

Writing an example expression for spatial dimension ≥ 3 per Appendix A

$\iiiiiiint dV = \iiiint dV(0)\ x\ \exp(r_V\ x\ t)$

where $r_V = \text{r-sub-V} = \varphi^{\wedge}(D(n+1) - D(n))$

then

$dx\ /dx(0) = \exp(\varphi\ ^{\wedge}\ (D(n+1) - D(n))^{\wedge}1/(n+1)$.

Except we are now doing math in another dimension, and while e = 2.718 in three dimensions, the base of natural logarithms should change in higher or lower dimensional space.

For example, in the case of 5 dimensions: e → e^(1/γ)^5/2.

We quickly find dx /dx(0) = 1.08.

A different example, for the case of spatial dimension < 3:

The base e must change as a function of the power of B, i.e. in three dimensional space B ~ sec^2 while in two-dimensional space B ~ sec^3/2.

The difference in power of physical events B

2 – 3/2 = 1/2 and the two-dimensional e = 2.718^1/2.

Then we quickly find dx /dx(0) = 1.08 similar to the previous mean calculations for the difference between b(min) and b-empirical.

The (Fibonacci) calculations hold true for any spatial dimension n moving through n+1 with a dimensional adjustment for e.

Appendix K (Reference)

The Various Sizes of Black Holes and Curvatures of Three-Dimensional Space

Appendix H defines the mass-radius relationship as observed in three-dimensional space:

$$m_H / r_H = K_G(\lambda)$$

where $\lambda = \Delta\lambda = C_{R3} / C_{R5}$

and represents the ratio of curvatures from 3-dimensional space and 5-dimensional space through 8-dimensional space respectively.

We assume, for the three dimensional intersections, that $C_{R5} = 1$.

The minimum $C_{R3} = 1 / c$ and the maximum $C_{R3} = c / c = 1$.

Allowed quantum are then n / c for $n = 1$ to c.

The minimum (least dense) intersection is an intersection among 3, 5 and 8 dimensional space where $C_{R5} = 1$ and $C_{R3} = C_{R3}(\min) = 1 / c$.

The next "largest" (more dense) intersection should occur for $C_{R3} = 2 / c$ and so on.

The most dense intersection occurs where $C_{R3} = c / c = 1$ and represents a closed third dimension in both eight dimensional and five dimensional space.

To visualize curvatures, the diameter of a circle = d is a straight line with curvature
$C_{R1-3} = 0$ while the circumference (length πd) closes upon itself (runs into the back of itself) and has the curvature $C_{R1-3} = 1$.

The curvature C_{R2-3} is closed in 3-dimensions visualized as a spherical (or elliptical, not reviewed in this scope) surface area that has closed itself around a center-of-mass c_M.

The two dimensional surface does not alter or "grow" in three dimensions, but the one dimensional line, e.g. the straight path of a distant comet or ray of light ($C_{R1-3} = 0$) or the line of a planetary satellite $C_{R1-3} = 1$) both curve (or bend) around mass in three dimensions to the two extreme degrees of curvature.

Then the ratio $m_H / r_H = K_G$ should represent a curvature of three-dimensional space through eight-dimensional space.

Appendix L (Reference)

Definition of Mass and Geometry for Black Holes

From the equation $E_B = a_G$ J kg$^\wedge$-1 $b_n$$^\wedge$-n x b_n (= 1.089E-16 on Earth surface,)

and from the definition of physical events in dimension n = B where B ~ sec$^\wedge$2 in 3-dimensions and sec$^\wedge$(D-1) per appendix C, then b_n has the following value:

3-dimensions: b_3 = 1.111E-17 meters (per the main text)
5-dimensions: b_5 = b_3 / c = 3.703E-26 meters
8-dimensions: b_8 = b_5 / c$^\wedge$2 = 4.114E-43 meters and so on.

Per Appendices G, H and K, the black hole geometry is a function of mass and becomes a series of symmetrically-closed concentric surfaces having internal densities:

8-dimensional volume in 3-dimensions = 4 / 3 π $r_8$$^\wedge$3
5-dimensional volume (= $\int$ (r_{5-8}-to-r_5) 4πr$^\wedge$2) = 4 / 3 π $r_5$$^\wedge$3 (including the volume of 8)
3-dimensional volume = 4 / 3 π $r_H$$^\wedge$3 where r_H = r_5 (including volumes of 5 and 8) and

the boundary condition for the most-dense black hole is then:

$$m_H / (4\pi r_H^2) = m_H / (4/3\ \pi r_H^3)$$

where $r_H = m_H / K_G$, then

$m_H(\text{max}) = 3K_G$ kg and the corresponding

$r_H = 3$ km.

The black hole mass m_H for the general case $r_H = r_S$:

$m_H(\lambda) = \int (\text{from } r_S - r_8 - \text{to} - r_S)\ 4\pi r^2 = 3K_G(\lambda)$

where

$\lambda = r8 / r5$,
$r_H = r_S$ and
r_8 is the 3-dimensional-radius of zero-mass 8-dimensional space ($b = b_8$) at the center of the hole.

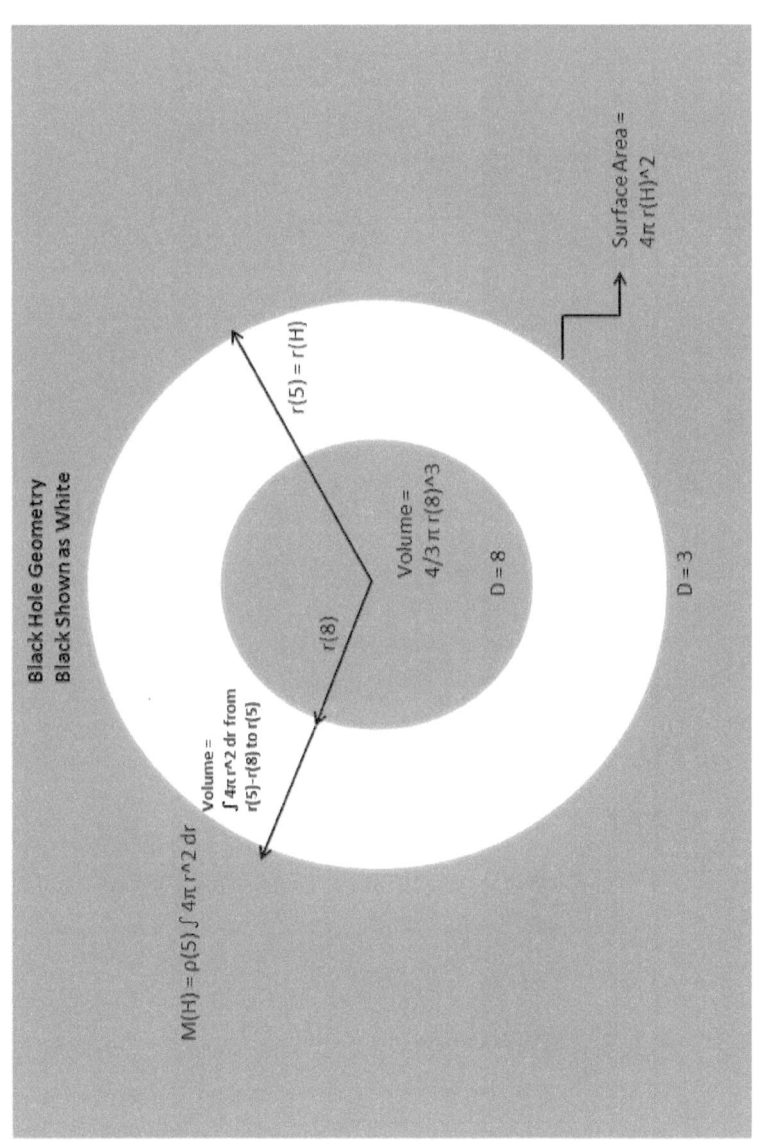

Black Hole Geometry
Black Shown as White

$M(H) = \rho(5) \int 4\pi r^2 \, dr$

Volume =
$\int 4\pi r^2 \, dr$ from
r(5)-r(8) to r(5)

r(8)

r(5) = r(H)

Volume =
$4/3 \, \pi \, r(8)^3$

D = 8

D = 3

Surface Area =
$4\pi \, r(H)^2$

Appendix N (Reference)

Nuclear Forces

Consider two possibly adjacent protons interacting through two 3-dimensional square law forces, the charge force and the mass (gravity) force. The forces act along straight lines (1-dimension) through square laws (2-dimensions) in 3-dimensional space.

For the gravity force, the magnitude is $G(1 \text{ amu})^2 / r^2$ where $r \sim b$.

Per appendix E, then mass x volume $= K_F / n2 \sim (1E\text{-}11)(1 \text{ amu})^2 \times 4/3 \, \pi \, (3/2 \text{ b})^3 / n^2$ or

Mass x volume $\sim (1E\text{-}11)(1E\text{-}27)^2(1E\text{-}17)^3 / n^2 \ll 1E\text{-}85$ for any n

where we have used the three-dimensional special relativity law mass $_{MIN} = E_B / c2$ to calculate the minimum bound on mass x volume $\sim 1E\text{-}85$ in a Joule system of measurement.

Therefore, the weaker of the two square law forces, by itself, breaks the allowed three-dimensional mass-volume rules for the extent of any nuclear size.

We suggest the atomic nucleus to be higher dimensional (b < 1.111E-17 meters per appendix L) having an exterior 2-dimensional-shell intersection with 3-dimensional space (volume.)

In that case, nuclear forces should not be bounded by one-dimensional square-laws; instead, they would be minimally bounded by two-dimensional cube-laws and should be relatively large.

References:

1. *Changing Your Mind, A Theory of Space without Time;* A Mathematical Transformation of Variables Defining Space-Time and the Constant h. Xlibris Press, July 2012. Marc E. King.
2. **Erwin Schrödinger**, Quantisierung als Eigenwert Problem, 1926.

3. **Leonhard Euler,** *Introductio in analysin infinitorium,* 1748.
4. **Ljunggren S.,** Journal of Magnetic Resonance 1983; 54:338.